I Wonder
Where Butterflies Go In Winter

• • • • • • • • • • • • • • • •

and other neat facts about insects

By Molly Marr
Illustrated by Paul Mirocha

A GOLDEN BOOK • NEW YORK
Western Publishing Company, Inc.
Racine, Wisconsin 53404

Produced by Graymont Enterprises, Inc., Norfolk, Connecticut
Producer: *Ruth Lerner Perle*
Design: *Michele Italiano-Perla*
Editorial consultant: *Steven J. Prchal*, Director, Sonoran Arthropod Studies, Inc.

Contents

What does it take to be an insect?

Flies, beetles, bees, wasps, ants, moths, mosquitoes, ladybugs, dragonflies, aphids, grasshoppers, butterflies, and crickets may look different from each other and behave in different ways, but they all have things in common that make them belong to a group called insects.

An insect has:

Three main parts to its body: head, thorax, and abdomen.

A pair of feelers, or *antennae,* at the top of the head.

Six legs, some of which can be specially developed for jumping, swimming, or grasping.

An outer covering, called an *exoskeleton,* that protects its soft insides.

Most insects have two pairs of wings. Flies have one pair. Some insects have no wings at all.

4

Why is a spider not an insect?

Spiders are not insects. They have only two body parts, not three. They have eight legs, not six. Spiders do not have wings.

Why is a hummingbird not an insect?

Some hummingbirds are not much bigger than bumblebees or moths, but they are not insects. They have just two legs and their bodies are covered with feathers. Unlike insects, they have a bony skeleton inside their bodies.

Why are there so many insects?

There are millions and millions of different kinds, or *species,* of insects—more than all the mammals, birds, reptiles, fish, and plants on earth. Most kinds of insects lay many eggs. The queen bee can lay two thousand eggs in just one day—and she continues every day for four or five years!

Amazing *but* TRUE

All the insects on earth combined weigh twelve times as much as all human beings!

Is a caterpillar a worm?

A caterpillar looks like a worm and moves like a worm, but it isn't a worm at all. Strange as it may seem, a caterpillar is what a butterfly or moth looks like before it is grown-up.

When most animal babies are born, they look like their parents. But when butterfly or moth babies hatch from their eggs, they don't resemble their parents at all. When they are young, their appearance keeps changing at different stages of their development. They don't look like their parents until they are adults themselves. These different stages of change are called *metamorphosis*. Most insects go through some form of metamorphosis.

Here's what happens:

1 The mother butterfly lays her eggs on the leaves of a plant that her babies will like to eat when they are born.

2 After a few days, little wriggly wormlike *larvae* hatch out of the eggs. The larvae of butterflies and moths are called caterpillars. The caterpillar's main job is to eat and grow. It starts by eating its eggshell and then chomps the leaves on which it was born.

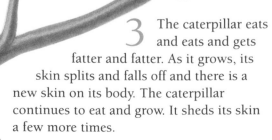

3 The caterpillar eats and eats and gets fatter and fatter. As it grows, its skin splits and falls off and there is a new skin on its body. The caterpillar continues to eat and grow. It sheds its skin a few more times.

4 When the caterpillar is fully grown, it stops eating and attaches its body to a leaf or twig. Then the caterpillar shakes its body so that its skin splits one final time. It forms a hard case called a *pupa*, or *chrysalis*.

6 When the chrysalis breaks open, a wrinkled little butterfly appears. It clings to a twig and pumps liquid from its body into its wings to expand them. This is the adult stage of metamorphosis.

5 Inside its chrysalis, the caterpillar develops wings and finally turns into a butterfly.

Amazing *but* TRUE

Some caterpillars can eat more than one hundred times their own weight in their lifetime. Lunar moth caterpillars eat close to their own weight each day. If you could do that, you'd have to eat about two hundred hamburgers every day!

7 As soon as the wings are dry and strong, the butterfly flutters away.

8 Now the adult butterfly darts from flower to flower. It does not munch on leaves anymore. Instead, it uses its long, hollow tongue like a straw to sip sweet liquid *nectar* from the center of flowers.

9 Soon the female butterfly will lay her eggs. Like her mother before her, she will not see her babies, because her life will be over before they are born.

Why do dragonflies have such big eyes?

You can tell by the dragonfly's large wings that it is a good flyer. Since it hunts for its prey while flying, it must depend on its eyes to spot passing insects. Luckily, dragonflies have the right eyes for that job.

How does a dragonfly hunt?

When a dragonfly spots an insect it wants to catch, it holds its spike-studded legs forward to form a kind of basket. Then it swoops the insect into the front opening of the basket. If the insect is lightweight, like a mosquito, the dragonfly will munch on its prey as it flies. If the insect is heavy, like a bumblebee, the dragonfly will be forced to land and have its picnic on the ground.

Now you see it, see it, see it . . .

Most insects have compound eyes. That means that when they look at something, they do not see just one image the way you do. Their eyes are made up of many tiny lenses. Each lens sees a separate image. Each image is a little different from the other, and all together, they make up one whole picture.

Since the dragonfly needs especially good vision in order to catch its prey in the air, it has huge, bulging eyes. Each eye has twenty-eight thousand lenses!

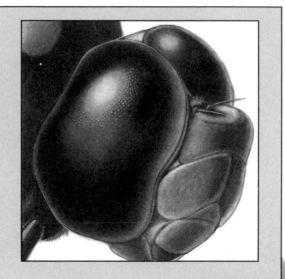

Tell Me More

Scientists believe that insects like dragonflies, which fly in the day, can see color. However, night-flying insects are blind to color and see everything in gray.

Amazing *but* TRUE

A dragonfly is one of the fastest flying insects. It can fly up to thirty-five miles per hour in still air. That's about as fast as a school bus travels. If the wind is blowing in the same direction as the dragonfly is going, it can fly even faster.

9

Why do fireflies flicker?

The pretty lights you see flickering during a summer night are male and female fireflies calling to each other.

There are hundreds of different kinds of fireflies and each flashes its own special signal or code that tells what kind of firefly it is. For example, some may send out one long flash, followed by two short flashes every few seconds. Others may just send out several short flashes.

Some female fireflies remain on the ground, flashing their light. The winged male flies above, looking for a signal that shares the same rhythm as his. When he sees such a signal light flashing from the ground, he knows he has found a female of his own kind. Then he flies right over.

Amazing
but TRUE

A firefly's light gives off no heat.

Tell Me More

Though a firefly is also called a glowworm or lightning bug, it is really a beetle. Some people say that a firefly is called a glowworm because some females have no wings and their bodies look like worms.

Where does the firefly's light come from?

The firefly's belly is called its lantern because a greenish light shines from a gland at its tip. But the light in the lantern comes neither from a light bulb nor from a flame. It is a strange, cold chemical glow that comes from inside the firefly's body.

Why do crickets chirp?

Insects can be recognized by the sounds they make. Bees buzz, katydids say *katy-did, katy-she-did*, but the most musical and beautiful sound is the cricket's sharp, sweet chirp and trill. In late fall, crickets all sing together in a chorus of short, low musical chirps in even beats.

Crickets have different kinds of songs. Some are love songs sung by males to attract females. One is a battle cry sung by male crickets when they are fighting. Another song is an alarm sounded to warn other crickets when danger is near.

Crickets change the speed of their chirps as the air temperature changes. The chirping is fast in warm weather and slow in colder weather.

The temperature cricket

The snowy tree cricket is also called the temperature cricket. That's because you can measure degrees in Fahrenheit by the number of chirps it makes.

Here's how: Count the number of chirps the snowy tree cricket makes in fifteen seconds. Then add forty to that number to get the Fahrenheit temperature! If a cricket chirps thirty-five times in fifteen seconds, add forty and you'll know the temperature is seventy-five degrees Fahrenheit.

How do crickets sing?

Crickets don't make sounds the way people do, because they have no vocal cords. And they have no ears with which to hear their own songs. The music comes from the male cricket's two wings, which he rubs together like a bow over violin strings. One wing is called a file. It has lots of rough little bumps on it. The other wing, called a scraper, has a hard bar on it.

When the male cricket serenades the female by rubbing his two wings together, the female cricket hears the song with "eardrums" that she has in her front legs!

Amazing but TRUE

Some people in China consider crickets good luck and often keep them as pets.

13

How do grasshoppers hop?

If you've ever snuck up on a grasshopper, you've seen its best defense—a quick, long leap away from you! Grasshoppers are such good jumpers because their two back legs are extralong and have muscles built especially for leaping.

A grasshopper can jump twenty times its own body length. If you could do that, you'd jump from the pitcher's mound to home plate in one hop!

Some grasshoppers have wings. They can go still farther. When they have jumped as high as they can, their wings start flapping to move them forward. When they want to slow down, their wings curve to catch the air. Then they drop gently to the ground again.

What is a katydid?

A katydid is a long-horned grasshopper that has very long antennae on top of its head. Its antennae are as long as its body.

A male's *katy-did*, *katy-she-did* sound can be heard by a female a mile away.

Why are many grasshoppers green?

Animals often have the same color as their surroundings to protect them from predators. The grasshopper's green color blends in with the leaves and grass where it spends most of its time. That makes it difficult to be seen by the birds that enjoy grasshoppers for dinner.

Tell Me More

Some female grasshoppers have long, thin tubes at the end of their bodies. They use these tubes to drill holes in the ground, plants, or trees. Then they lay their eggs in the holes.

When the young grasshoppers hatch from the eggs, they don't go through metamorphosis the way butterflies and many other insects do.

15

How can a fly walk upside down?

Common houseflies can do some uncommon things. Their feet are equipped for climbing on all kinds of surfaces. The tips of their feet have hooklike claws, which they use to grab on to uneven surfaces like wood-paneled or stone walls. Below the claws, two pads covered with a sticky substance keep the flies from falling off smooth walls, windows, and even ceilings.

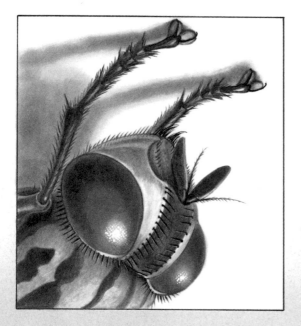

Shoo, fly!

The housefly's sticky feet are wonderful for walking up walls, but they create a problem for people. These sticky feet may pick up germ-laden particles from garbage dumps or other unclean places and deposit them on the next thing they land on. If that happens to be your lunch, the germs could make you sick. That's why food should be covered—especially in summer.

Tell Me More

Its feet aren't the only surprising parts of a fly's body. Its feelers, or antennae, act like a nose—for smelling. The antennae can detect odors at long distances.

Fabulous feet

Believe it or not, a fly's feet are also used for tasting. Because its mouth is shaped like a sponge, it can eat only juicy or soupy foods. So when the fly lands on a piece of food, its legs tell whether the food is liquid and ready to eat, or solid and in need of mashing and tenderizing.

When its feet land on something liquid, such as melted ice cream, they signal the fly to open its mouthparts and suck up the food. When they land on solid food, such as a cookie crumb, they signal the fly's stomach to produce special fluids that the fly spits out to soften the food before it can be sucked up.

View from the top

A fly has two big compound eyes, each with thousands of lenses. But that's not all. Hidden on the top of its head are three "simple," or single-lens, eyes, which can be seen only under a magnifying glass.

Amazing *but* TRUE

Houseflies spend their whole life in the area where they were born. Male houseflies live only about two weeks, and females a month. In this short time, females can produce five thousand eggs!

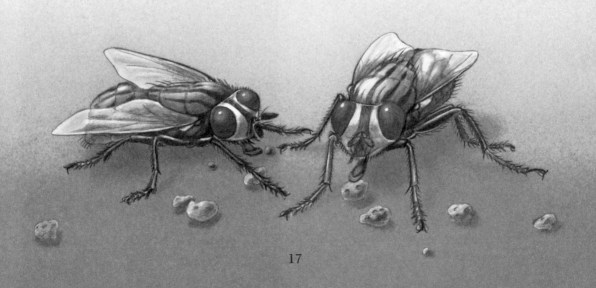

What's an anthill?

Ants collecting seeds and carrying them into the nest

The little mounds of sand or earth you see along the ground when you're in the country are called anthills. These little hills are the sand or earth that ants have scooped out in digging their underground nests.

Ants are social insects. So are most bees and wasps. That means they work together in a community, or colony, just as people do. There are different types of ants in a colony, and each has a special job to do.

Are ants dirty?

Ants don't use soap and water, but they are very particular about being neat and clean. The ant has a comb in the middle of each front leg. It uses this comb to remove dirt from its antennae and hind legs. Then it cleans the comb with its mouth. Worker ants groom the queen and young ants.

Ant digging new tunnels

Different ants build different kinds of nests. These are seed harvester ants.

Workers cleaning queen and her chamber

Queen laying eggs

18

Ant carrying trash out of the nest

Nest entrance or vestibule

Do ants keep cows?

Some ants tend herds of tiny insects called aphids, which they keep in special compartments in the nest the way dairy farmers keep cows in barns. When an ant strokes an aphid's body, the aphid releases droplets called *honeydew*. Worker ants collect these droplets and feed them to the queen.

Ants caring for eggs and pupae

Why do ants come to a picnic?

If you drop a sandwich crumb on your picnic blanket, chances are you'll soon see an ant . . . then, two . . . three . . . four . . . and more! That's because ants, like other social insects, work in groups. When an ant finds a crumb, it lets the others in its colony know there's a treat to be had.

On its way back to the nest, the ant leaves a trail of a special scent. Then the other ants follow the scent trail to where the first ant found the food.

Amazing but TRUE

If you rub your finger across the ants' path, the line of ants will stop. That's because you rubbed out the smell that was leading them to the food.

What if a crumb is too big for one ant?

When a tasty morsel is too big for one ant to carry, the ants try to break the food into smaller pieces. But if that can't be done, they don't give up. They lift and carry their load together. It takes a group of ants about ten minutes to figure out a way to do it, but they always do!

Are there male ladybugs?

Despite their name, not all ladybugs are female. Like most other living things, ladybugs need both males and females to reproduce. Both male and female ladybugs can be recognized by their black-spotted bright-red wing covers. They may seem identical to us, but not to other ladybugs. They cannot tell by sight, but by smell! Males and females have different scents.

Why are ladybugs a gardener's friend?

Even though they are smaller than most other beetles, ladybugs have giant appetites. Their favorite meal consists of tiny plant-eating insects called aphids, or plant lice. That's good news for farmers and gardeners, since aphids destroy crops.

Ladybugs are powerful hunters and killers equipped with sharp jaws and strong claws called pincers. Believe it or not, one ladybug can eat one hundred aphids a day!

Tell Me More

Gardeners who don't want to use chemical insecticides to kill aphids buy ladybugs by the pound and then let them loose in the garden.

How does a wasp make paper?

Some wasps, including hornets, knew how to make paper out of wood long before people ever did. They were the first manufacturers of paper on earth. And they didn't need machinery or special tools. That's because wasps have most of what they need right in their own bodies.

Wasps use the paper they make to build their nests. These nests are so well built that they last through rain and wind.

Why does a wasp have a comb?

The wasp's nest is called a comb. It has many rooms, or chambers, inside. The queen wasp lays one egg in each of the chambers. These eggs hatch and develop into workers that will make more paper and build more rooms. The queen keeps laying more eggs. The more eggs she lays, the more workers there will be to make paper and enlarge the nest.

Do hornets really get mad?

Nothing seems to make hornets angrier than someone who disturbs their nest. If an intruder comes too close to the nest, they zoom straight toward the person or animal and sting. Anyone who has been stung knows how "mad" hornets can get!

Amazing *but* TRUE

Besides building their comb, worker wasps must also hunt for food for the young wasps. They kill caterpillars and bring them back to the nest.

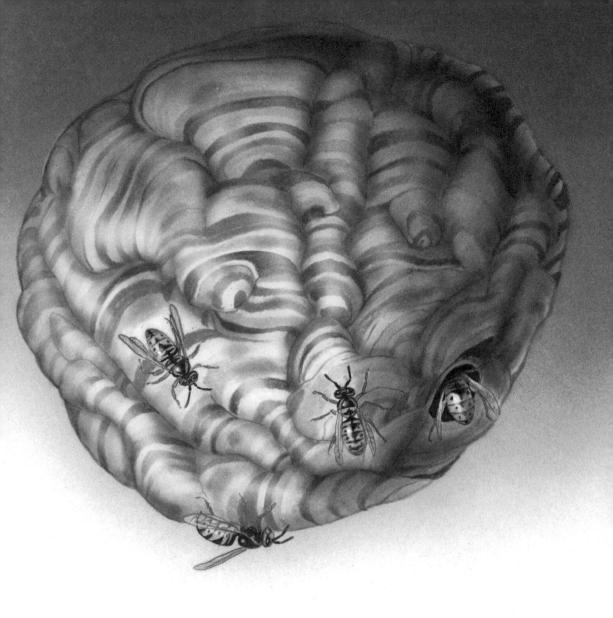

Does the queen wasp wear a crown?

The queen wasp does not wear a crown, but she is very special in other ways. She is the only member of the wasp colony to survive the cold winter. When it is time to build a new nest in the spring, she must start building all by herself.

First the queen chooses the place for her nest, which is usually on the branch of a tree or under a roof. Using her mouthparts, she scrapes wood from dead trees and old fences. She chews the wood and mixes it with her saliva. This makes a kind of paste that can be stretched into thin strips of paper. The queen uses the paper strips to shape the nest.

How do spiders spin webs?

Insects have many enemies, but none sets a trap as clever as the spider's. Some spiders use webs to catch their meal of insects, just as fishermen use nets to catch fish for dinner.

Glands in the spider's abdomen make a silky liquid that is forced out of its body in thin streams. When this liquid comes in contact with the air, it hardens into a delicate but strong thread that the spider uses to make its web. Each species of spider weaves a web in its own special pattern. It takes about an hour to make a web, and most spiders weave a new one every single day.

"Will you walk into my parlor?" said the spider to the fly.

When a fly gets caught in the sticky threads of the web, it struggles to get free. This struggle makes the web vibrate. By feeling this movement, the spider can tell exactly where its victim is entangled. Since most web-building spiders have poor eyesight, they must find their prey by following the web's vibration. Then they wrap their victim in a silk thread and carry it back to the center of the web, where they enjoy their dinner while waiting for dessert to come flying by!

Amazing but TRUE

Baby spiders can weave a perfect web on their very first try!

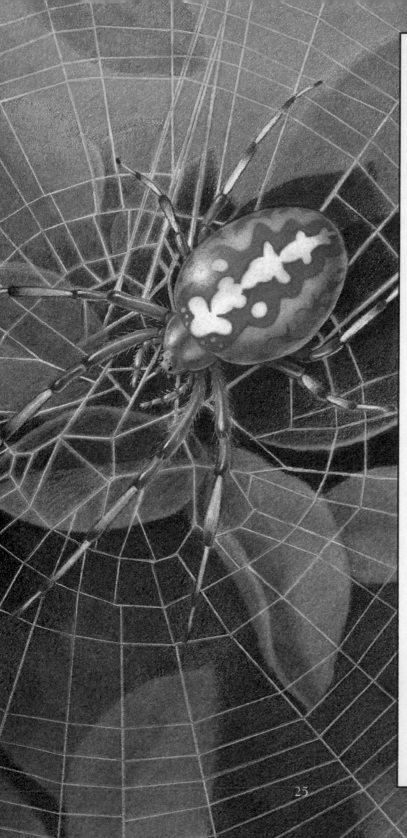

Here's how the orb-spider builds its trap:

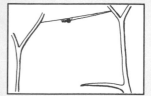

1. The spider anchors its web.

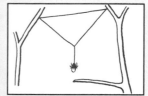

2. It forms the outside.

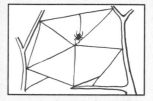

3. It spins threads that look like the spokes of a wheel.

4. It adds more threads at the center to strengthen the web.

5. It weaves sticky threads into a spiral around the edge of the web and into the center.

Why do  moths fly toward light?

Unlike butterflies, which fly by day and rest by night, moths fly during the night. Since it is hard for moths to see in the dark, they use the moon to guide them as they fly along. The light of the moon acts like a lantern in the sky, showing them the way.

Moths fly toward the light by instinct, so they will fly toward any light, even if it is not the moon. When the light is nearby, the little creatures fly straight into it!

How do moths make moth holes?

If you see some small holes in a wool sweater or hat, chances are they were made by clothes moths in the caterpillar stage. Clothes moths lay eggs on wool or fur so that when their young hatch, they'll have something good to eat—like the sleeve of your favorite sweater. Adult moths do not eat wool.

What's the difference between a moth and a butterfly?

- When a moth is at rest, it keeps its wings folded along its body.
- Most moths have chubby bodies.
- Most moths have feathery antennae.
- Most moths fly at night.

- At rest, a butterfly's wings are together and held straight above its body.
- Butterflies have slender bodies.
- Butterflies have smooth antennae with knobs at the end.
- Butterflies fly in the daytime.

Amazing *but* TRUE

The Atlas moth, one of the largest moths, can measure as much as twelve inches from wing to wing.

How do bees make honey?

Bees manufacture one of the tastiest treats on earth—that sweet, thick golden syrup called honey. They store the honey in little wax *cells*, which are compartments that make up the *honeycomb*.

Bees are social insects. Each has a special job to do in the *hive*, or nest.

Worker bees who work in the fields suck up nectar from flowers and carry it back to their hive in a special *honey sac*, which is located in front of the stomach.

Worker bees inside the hive take the nectar and fan their wings to dry up the excess water. Then they add their digestive juices to turn the nectar into honey.

The bees store the honey in honeycombs for future use. The bees' wax-making glands make wax for the honeycomb cells.

Why do bees need to make honey?

Squirrels collect and hide nuts and acorns that they will eat in winter. Bees make their own food—honey—to tide them over the cold months. In winter, when the bees huddle together in their hives, they open the honey cells and feast on the honey they made in the summer months.

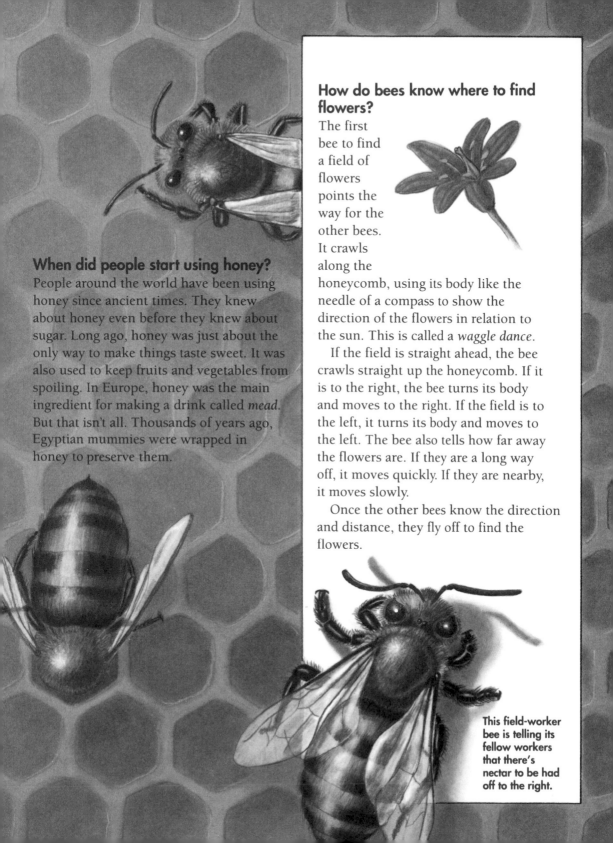

When did people start using honey?

People around the world have been using honey since ancient times. They knew about honey even before they knew about sugar. Long ago, honey was just about the only way to make things taste sweet. It was also used to keep fruits and vegetables from spoiling. In Europe, honey was the main ingredient for making a drink called *mead*. But that isn't all. Thousands of years ago, Egyptian mummies were wrapped in honey to preserve them.

How do bees know where to find flowers?

The first bee to find a field of flowers points the way for the other bees. It crawls along the honeycomb, using its body like the needle of a compass to show the direction of the flowers in relation to the sun. This is called a *waggle dance*.

If the field is straight ahead, the bee crawls straight up the honeycomb. If it is to the right, the bee turns its body and moves to the right. If the field is to the left, it turns its body and moves to the left. The bee also tells how far away the flowers are. If they are a long way off, it moves quickly. If they are nearby, it moves slowly.

Once the other bees know the direction and distance, they fly off to find the flowers.

This field-worker bee is telling its fellow workers that there's nectar to be had off to the right.

Where do butterflies go in winter?

Insects can't wear sweaters to keep warm in winter and they don't have furry coats. They are cold-blooded, so their body temperature is the same as the air around them.

Many butterflies freeze and die when the weather gets cold. Some, like the monarch butterfly, keep from freezing by traveling to faraway warmer places. They go to the same place and take the same route each year. This kind of traveling is called *migration*.

Amazing but TRUE

The monarch butterfly loves to feed on the milkweed plant. The milkweed plant contains a substance that causes the monarch itself to have a bitter taste, and so its enemies don't enjoy eating it. Any bird that has tasted a monarch knows to stay away when it sees this butterfly's markings.

Where do monarchs migrate?

One of the most amazing migrations is that of the monarch butterfly. All the monarchs from the northeastern United States and Canada travel south together to a small area around the Sierra Madre Mountains of Mexico. A monarch butterfly that starts out in New Hope, Pennsylvania, has to fly some twenty-five hundred miles or more to reach its destination.

When the monarchs finally arrive, every inch of ground, every bush, every tree, every meadow, and every mountainside is covered with orange and black butterflies. Then the monarchs are ready for a long rest. They stop eating, they breathe very slowly, and they do a lot of sleeping.

What happens to monarch butterflies after the winter?

In the spring, the monarchs start on their way north, but this time they are not in a hurry. On the way, the female stops to lay her eggs on milkweed leaves—the monarch's favorite food. After all her eggs are laid, the female monarch dies. The young monarchs that hatch never get to meet their parents. They go through metamorphosis and then continue their journey north.

Tell Me More

Though you have come to the last page of this book, you are only beginning to know about the wonderful true-life stories of insects. Scientists who study insects are called *entomologists*. But you don't have to be an entomologist to enjoy finding out more about these amazing members of the animal kingdom.

There seems to be a plan and a purpose for everything in nature. Large or small, beautiful or strange, each plant and animal has a role to fulfill. Each has an effect on something else that sooner or later has an effect on us.

Here are some more amazing-but-true facts to start you on your way to new discoveries:

- One bee can call hundreds of bees to come to defend the hive by releasing a special scent from its body.

- Atta ants tend their own underground gardens in which they grow a kind of fungus that feeds the whole colony.

- Mayflies are adults for just one day. Since they must mate, lay eggs, and die within just a few hours, they have no time to eat, and have no mouthparts.

- If insects disappeared from earth, so would many birds and freshwater fish that depend on insects for their food.

- The tiny flea can do a broad jump about three hundred times the length of its body.

- The prairie sphinx moth was thought to be extinct, or gone from the earth forever. It was discovered again in Colorado about twenty years ago.